CONTENTS

Summary

This book is for you and your dysfunctional nose. It will help you better understand the sensitive fella and for some people, this may even establish an emotional bond. For starters, have you ever found that it was irritatingly harder to breathe through one nostril instead of the other? The reason why may not be what you expect! We shall discuss what a nose is, why you even have a nose and how it does its job. Following that, we will help you to understand the trademarks of "nose problems", running and blocked noses. Including how and why it happens. Doesn't mucus seem unpleasant and weird... What do we even need that stuff all up in your nose for? The functions, positive qualities and even the colour of your mucus will be explained, as well as how it may make things worse if you don't clear it! If you face "sinus" based problems in the morning or late at night, a chapter on environmental triggers will prove to be an especially good read. To end off the book, we decided to test, prove and rate certain ways you can combat your nose problems. A nicely curated selection of solutions, based on scientific facts and figures, will be presented with all the procedures and materials for you to follow and help you get some much-needed relief! Together, we can win the war against your nasty nose problems and make nasal discomfort a thing of the past!

DEFINITIONS

Elasticity
- The ability of a deformed material body to return to its original shape and size when the forces causing the deformation are removed; Stretchiness.

Engorged
- Swelling caused by fluid build-up.

Glycoproteins
- "Glyco" refers to the presence of sugar. Thus, a glycoprotein is a protein molecule with an attached sugar molecule.

Inflammation
- Is a physical condition caused by the immune system which results in swelling of body parts.

Infection
- The invasion of pathogens in tissues.

Ions
- Charged particles.

Nasal Patency
- Is how wide the cavity in your nose is.

Pathogens
- Is a bacterium, virus or other microorganisms that can cause disease.

Olfaction
- The action or capacity of smelling; the sense of smell.

Mucopurulent
- Encompasses containing both mucus and pus.

Nasal Epithelial Cilia
- Refers to the short microscopic hairlike vibrating structure

found in large numbers on the surface of nasal cavity epithelial cells, causing currents in the surrounding fluid.

Nasal ulceration
-    Deep erosion in the nasal cavity exposing nasal cartilage.

Neuropeptides
-    Small protein-like molecules (peptides) used by neurons to communicate with the nervous systems.

Normal flora
-    Bacteria found in or on our bodies on a semi-permanent basis without causing disease.

Rhinitis
-    Is the inflammation of the mucous membrane of the nose.

Serotype
-    A distinct variation within a species of microorganism or immune cell. Defined and grouped by cell surface molecules.

Viscosity
-    Is the measurement of resistance in the fluid to change its shape.

PROLOGUE

Ever had a cold on a workday? Maybe it was just one of those days, perhaps a lack of sleep. Your nose ran like the wind, nasal fluid on the verge of coming out, only to be stopped by constant sniffles throughout the day. Tissues pile up on the desk as your supply of tissue packets in your bag run out!

This isn't your typical science book. 'Just being nosy' is a book written by two dudes with terrible nose problems for your terrible nose problem.

Ambrose can sneeze 21 times in a row, almost suffocating himself in the process. While Lucas's nose can run like a waterfall in the day and get blocked up like a dam at night. By the way, this book was mostly written at night, with one author sneezing away and the other being only able to breathe through his mouth.

If you have anything close to what we experience, this book is an absolute must-have. We'll go through "Nose Basics", bring you through what condition you might have and may even give you a break from all those years of discomfort!

CHAPTER 1: INTRODUCTION

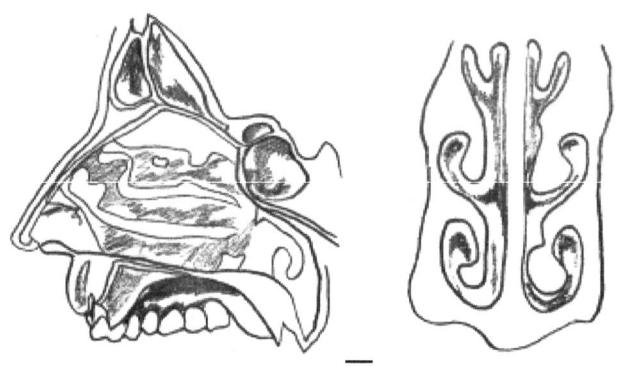

Your nose is an essential part of your respiratory system. It helps to defend you from nasty pathogens, produce sound and allows you to smell. It even warms and moistens the air you breathe for your lungs' comfort!

Its structure is separated into two parts, "What you see" is the external nose, while "What you don't see" being the internal nose. The latter is more important in figuring out what happens up there.

The internal nose

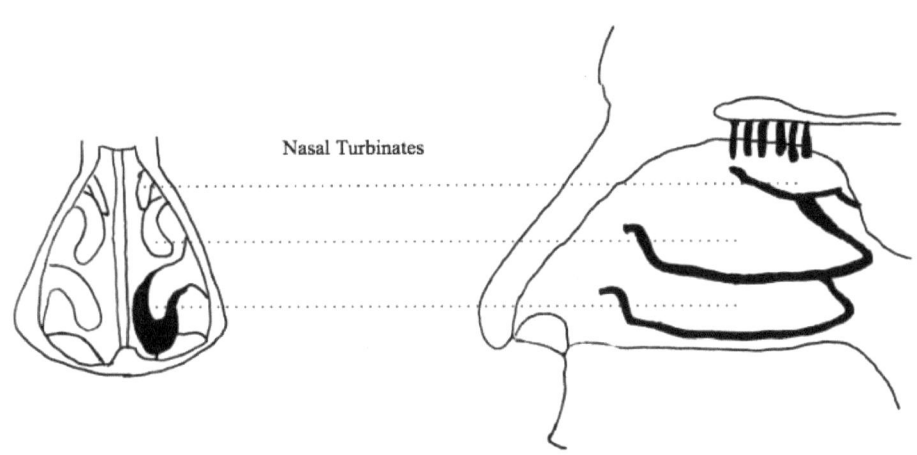

Nasal Turbinates

This is the inside of your nose; Isn't it much deeper than you expected? It goes in about 10-12 cm! (Mygind & Änggård, 1984, 173) Made of nasal cartilaginous tissue, connecting to the nasal bone, it acts as a channel for air to pass through, cavity walls are lined with blood vessels and sets of natural filtering mechanisms.

The Nasal Cycle

When your nose gets congested, have you ever felt that it was easier to breathe through one of your nostrils? Matter of fact, your nostrils take turns "opening and closing" every 1.5 - 4 hours when you're awake! This "switching-sides" phenomenon is described as the 'Nasal Cycle'. Physically it is controlled by the increase of blood flow to certain parts of the nose, the Inferior Nasal Turbinate and the Nasal Septum, causing periodic blood vessel dilation. The tissue swells up and narrows the nasal passage on one side.

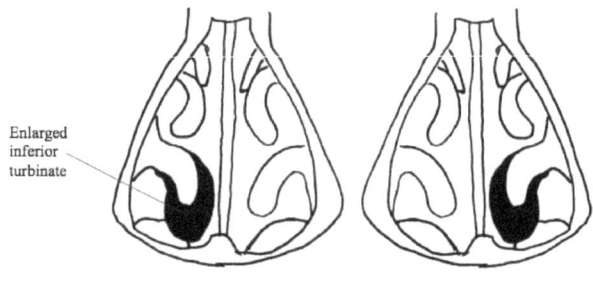

Enlarged inferior turbinate

"Switching-sides" phenomenon

During nasal congestion, the nasal cycle inflammation is exaggerated making it more pronounced and noticeable. Narrowing occurs in the nasal passage on the engorged side of the nose. As such, we experience congestion on one side of the nose.

So why does your nose undergo the nasal cycle or even have 2 nostrils in the first place? You may think it unnecessary, but there is more than meets the eye.

Olfaction

Smelling occurs when an odour particle in the air enters the nose and binds with receptors, sending signals to the brain. Different types of receptors are found in different parts of the nose, each type sending different smell signals.

Some odour particles are considered more 'sticky' to certain olfactory receptors, running the risk of not being fully detected ("smelled") if it doesn't reach the receptors at the back of the nose. Thus, a faster airflow, caused by the increased pressure, allows the odour particles to bind to a much larger range of olfactory receptors, even those at the back. A good example of this is "sniffing" the air when trying to smell, this creates a temporary increase in airflow speed, allowing you to fully and accurately comprehend the specific odour. In the nasal cycle, the more engorged side of the nose performs this role. (Kahana-Zweig et al., 2016)

On the other hand, the less 'sticky' odour particles need more time to bind to the receptors. Thus, the slower airflow, caused by the decreased pressure, allows the odour particles to properly bind to olfactory receptors found at the beginning of the nose, to once again produce a more accurate and pronounced smell signal. In the nasal cycle, the less engorged side of the nose performs this role. (Kahana-Zweig et al., 2016)

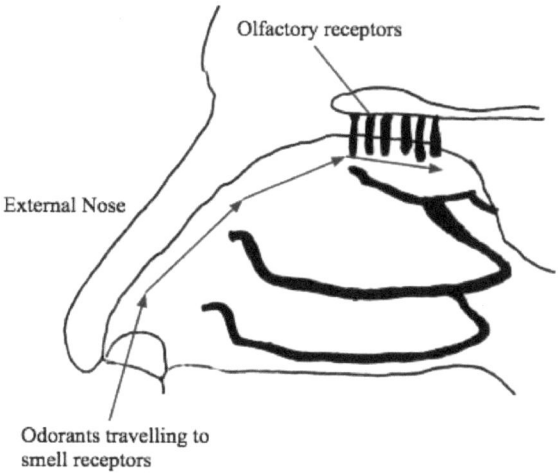

As certain smells can only be fully picked up at certain minimum airspeeds, narrowing one nostril to make airflow faster through the other is beneficial when it comes to smell, giving us a wider olfactory range. (Kahana-Zweig et al., 2016) Scientists speculate that the nasal cycle is also involved in the removal of debris from nostrils and the maintenance of Nasal Epithelial Cilia by keeping it moist and protected from dry air once in a while.

The nasal cycle further improves your sense of smell and shares the workload of the nostrils.

CHAPTER 2: BLOCKED AND RUNNING NOSE

A blocked nose put into a simple equation looks like this:

Running Nose = Excess mucus production + Condensation
Blocked Nose = Inflammation + Excess mucus production

Blocked nose are often a result of by viral infections. It is described as 'Infectious Rhinitis' or upper respiratory viral infection (URI).

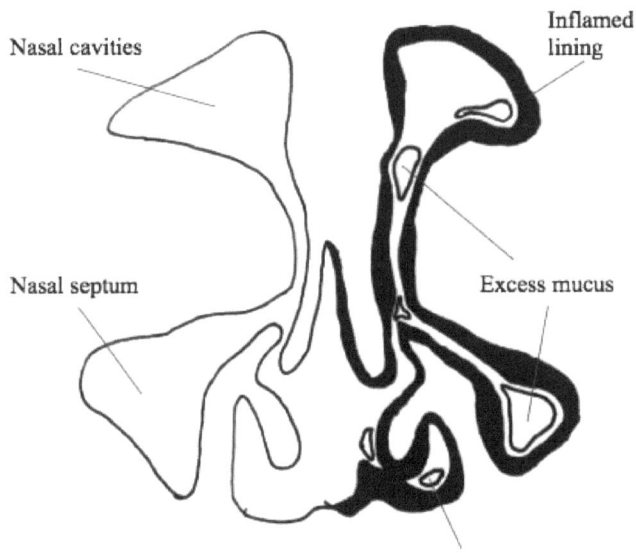

Nasal cavities

Inflamed lining

Nasal septum

Excess mucus

Possible site of infection

It usually affects the nose by causing inflammation (your body fighting back). Your blood vessels swell, blocking the nasal passages and stimulating mucus glands in the nose. In addition to a running nose, it'll typically cause a blocked nose.

Other common symptoms include mild fevers, pain and pressure behind the face, a scratchy sore throat and coughing. (Corey, Houser & Ng, 2000)

Should you be concerned about it? Well, thankfully infectious Rhinitis usually loses the battle with your body within a week. However, do take note that a less common 'Bacterial Rhinitis' may persist slightly over a week and even cause more severe symptoms. Less common, more serious, symptoms may include, headaches, body aches, itchy eyes or even a loss of smell in prolonged experiences with Rhinitis.

Rhinitis can also be caused by allergies; It can be affected by seasonal or environmental changes such as humidity and temperature. Personal and family history of allergy to medicines or other allergies can also act as potential stimulants. (Stewart, Peden, Thompson & Ludwig, 2012)

Mucopurulent Rhinorrhea (A symptom of Rhinitis)

Before you know it, stacks and stacks of tissue start piling up... You have a running nose. What you have is called rhinorrhea. Rhinorrhea causes excessive production of mucus, causing your nose to 'run'. When paired with inflammation, excessive mucus contributes to the clogging up your nasal passage, blocking air from passing through the nose. As aforementioned, this combination may also result in a blocked nose.

Rhinorrhea is caused by an infection of viruses. These viruses may also be able to develop into common colds (Viral Nasopharyngitis) or an upper respiratory infection if left untreated.

Like a chain-reaction, this leads to normal flora in the nose becoming opportunistic. This is how the story goes:

You're on your way to work and pass by some guy who sneezes in your face, sending 40000 droplets of pathogenic matter into the air.

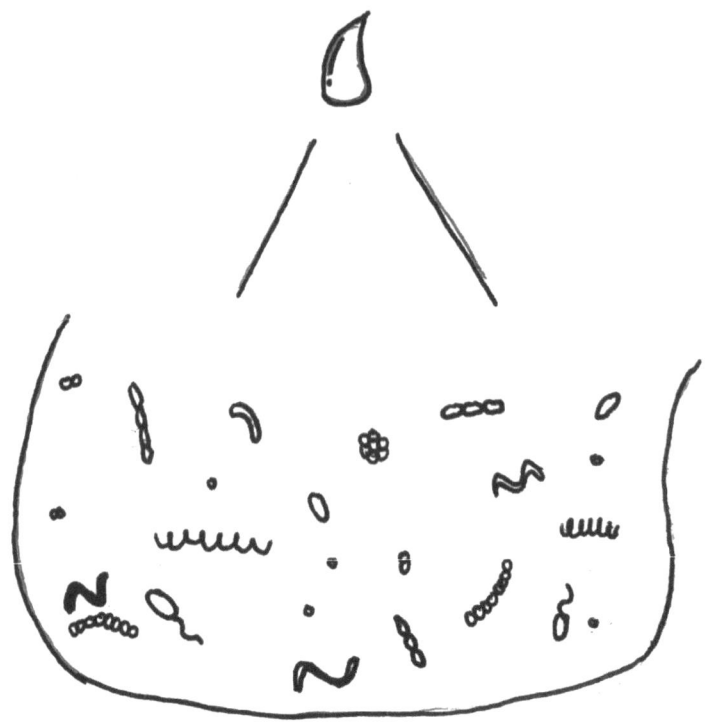

A single droplet of mucus from a sneeze can contain a lot of nasty bacteria, viruses and other pathogenic matter. Sneezes can travel at speeds of up to 100km/hour and linger for within 0.7m of the sneeze direction. (Bourouiba, Dehandschoewercker & Bush, 2014)

You breathe it in and happen to get infected by a virus. Your body's immune system gets activated and starts eating up the virus-infected cells. However, this virus may be new, one that your body has not faced before! Or maybe even a sneaky pathogen you have encountered previously that has changed its appearance?

Look! A single pathogen can appear in different forms! By simply changing its outward appearance, like the person donning an extra accessory, it can disguise and fool your body's immune system. The immune components will no longer be able to recognise and respond even though they are innately the same. As such, it takes time and struggles to kill them all off.

While the body is at war with the new virus you've contracted, usually beneficial bacteria in your nose (Most commonly Streptococcus pneumoniae) starts seeing opportunities to infect and take advantage of the body. Internal walls of cavities of the respiratory tract are left defenceless and S. Pneumoniae then starts a secondary infection at the nose. This causes an immune response from the body, inflammation occurs, over stimulating the mucus glands to produce more mucus, causing rhinorrhea. (Ferrándiz, Martín-Galiano, Schvartzman & de la Campa, 2010)

It results in the overproduction of thickened, opaque nasal discharge. Initially clear, nasal discharge can become white, yellow or green due to what secretions are present in the mucus. (We will dive deeper into this in the next chapter '**Mucus**') With the overproduction of this thick mucus, your nose gets all stuffy and airflow gets blocked.

CHAPTER 3: MUCUS

What is mucus? Generally, mucus serves as a protective coating of the respiratory tract. It can be found throughout the respiratory tract, from your nose to the insides of your lungs. Mucus produced by your body is made out of proteins, such as various Glycoproteins and Mucin , DNA and ionic compounds. (Padrid & Church, 2008)

Fun fact: Typically, mucus produced by your body starts in a dehydrated state, the presence of moisture in the nose moisten the mucus, allowing it to form a viscous and elastic mixture. In a mere 2 seconds, dehydrated mucus produced absorbs water in the surrounding and increase in volume, actually by a few hundred times! (Quraishi, Jones & Mason, 1998)

Viscosity and elasticity

Normally, healthy mucus is moderately elastic and thick but fluid, this is caused by a protein you naturally produce, muco-proteins! It gives the mucus it's properties but balances it with fluidity. However, when you get a cold, the thickness of your mucus is on a whole new level.

When experiencing a sickness, don't you find that your nose's mucus stuck in your nose? No matter how hard you blow your nose, your mucus refuses to leave. Tissues are literally everywhere, non-stop sniffling and you can't seem to clear your nose. Well, when your body fights a pathogen there's a lot of debris, just like in a real battlefield! Among the debris, DNA from both your cells and the pathogenic material is discharged. It might be easier to imagine it as a bunch of dead bodies being piled up.

DNA from the cell bodies in the mucus plays a major role in affecting the viscosity (thickness and stickiness) as well! As such when your body is fighting an invasion of pathogens (an infection), the number of dead cells start to pile up in the nose. The dead cells start mixing with the mucus and your mucus gets all gooey and sticky, clogging up in the nose.

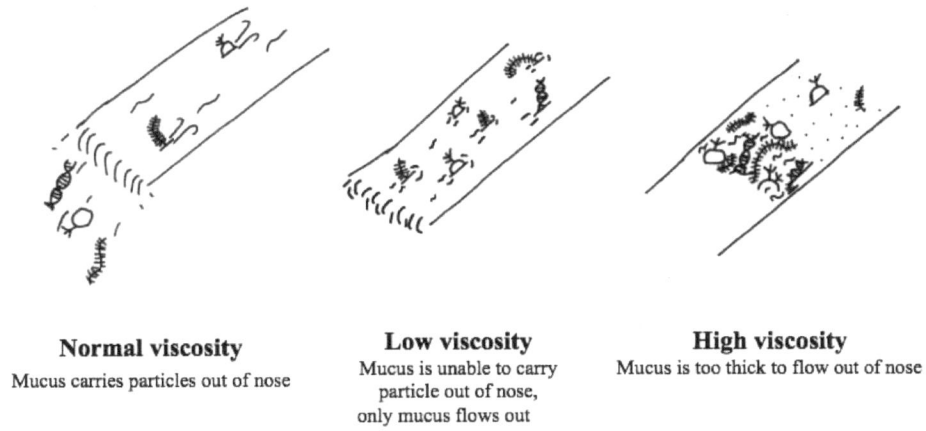

Normal viscosity
Mucus carries particles out of nose

Low viscosity
Mucus is unable to carry
particle out of nose,
only mucus flows out

High viscosity
Mucus is too thick to flow out of nose

The mucus has to maintain a balance for both viscosity and elasticity. If the mucus is too elastic, the mucus would be like a dough, unable to flow. If it is not elastic enough, the mucus will separate and not flow in unicen.

If the mucus is not viscous enough, the mucus will not be able to support the particles. Thus, mucus would flow around and pass the particles. This will cause the particles to remain in the nose, resulting in a particle build-up. If it is too viscous, the mucus would be like dough. The cilia in the nose will encounter too much resistance to push mucus out. The mucus will be able to support the particles but will be unable to flow as it is too thick for the cilla to transport. (Eylers, 2001) This also results in a build-up of particles. Furthermore, as particles build-up, a snow-ball effect on viscosity will also be caused.

A constituent protein of mucus, such as Mucin, exhibit carbohydrate properties (glycoproteins are a combination of carbohydrate and protein). As bacteria is capable to digest mucin, if the mucin is unable to be discharged, bacteria will take advantage of the clogged mucus as nutrition to grow! Hence, clogged up mucus may serve as a breeding ground for bacteria, where they are allowed to flourish. The once faithful protective mechanism against pathogens, backfires when mucus is overly viscous and becomes a free "power-up" to further strengthen enemy pathogen's prowess, this further aggravates the infection. (Carrington, Clyne, Reid, FitzPatrick & Corfield, 2010) As such, a sneeze may turn much more infectious than normal. Under the influence of an infection or cold, a single droplet of mucus sneezed into the air may contain millions of pathogenic bacteria!

Colour of Mucus

The colour of mucus also has the potential of giving some hints of what is going on in your nasal passages and body. Feel free to blow your nose and evaluate it as you go along!

Clear - You are doing A-OK
Mucus is healthy, everything is balanced. Your nasal tissues produce this stuff all day, every day. Most of it flows down the throat and is dissolved in the stomach.

White - Having a blocked nose?
Your mucus might be a bit dehydrated, causing it to be particularly thick and cloudy. A probable cause would be inflamed nasal tissue, slowing down the flow-rate of mucus. This might be an early warning that you have contracted a cold or nasal infection.

Yellow - The cold is starting to develop!
Early defences are triggered, white blood cells respond to the infection and start fighting the pathogens. Once they have done their jobs, they are then discharged in the mucus, causing the yellowish colouration. Colds inevitably last 10-14 day, it's best to take it easy and wait it out.

Green - Your body is fighting at full force
Mucus is thick and dead cell debris is accumulating. Your infection is squaring up to be quite the opponent. If your condition shows no improvement for more than 12 days, you should consider getting it checked by a doctor. Especially if you are experiencing any other symptoms.

Pink/Red - Blood 0.0
There is blood in your mucus! Maybe you had a bleeding nose last night? Dryness from a lack of moisture, irritation and impacts are possible causes of damaged nasal tissue. The presence of blood indicates abrasion and bleeding, caused by violent sneezing or

nasal ulceration.

Brown - Dirt, could be blood.
Maybe it was some spices you breathed in earlier, possibly dirt, debris and maybe dried blood.

Black - Something is up! Go see a doctor.
Experiencing black mucus, while not being in the habit of smoking or illegal substances, may depict an immune system in trouble! This infection may be fungal related. Seeing a doctor is an absolute priority. ("What the Color of Your Snot Really Means", 2017)

CHAPTER 4: ENVIRONMENTAL FACTORS AFFECTING THE NOSE

Usually, when you are on your way to school or work, does your nose suddenly starts stuffing up and ruining your morning? That's vasomotor rhinitis or "sinus" which everyone casually blames it on. Vasomotor rhinitis is non-allergic rhinitis that has chronic (persistent and regular) symptoms of nasal congestion or rhinorrhea. It may even worsen due to cold air, haze and specific emotional upset without clinical symptoms (fever or sore throat). (Pappas, 2018) This is due to some people having nasal receptors that are more sensitive than others which leads to the frequent occurrence of vasomotor rhinitis or "leaky" noses at certain times of the day. (Fokkens, Hellings & Segboer, 2016)

Humidity

Humidity also directly affects the mucous membranes of your respiratory system and produce adverse effects on your health. The humidity levels should always be kept within a certain range. Ideally between 40-70%.

Infections are on a case-by-case basis and usually are a result of a combination of factors. However, it is important to note that low environmental humidity, under 40%, gives a higher chance of coming down with upper respiratory infections. Based on where you stay in the world, you may want to consider using humidifiers in the house to stay above 40% humidity, this will provide a more comfortable condition for your nose, reducing the chances of infections. Singapore's humidity fluctuates between 64-96% and has an average humidity usually about 80%. As such you do not have to worry about humidifying your home regularly.

That being said, having an environment with too high humidity can also cause some problems. For instance, when the humidity is above 70%, the survival rate of some airborne-transmitted viruses may increase. An increased spread of pathogens may not only further pass on the virus to others but also exacerbate the condition of a person with an already compromised immune system, introducing more pathogens into the system. As such it is recommended to reduce the humidity of a room below 70%, especially when someone around you is not feeling well. High humidity can be lowered with the use of dehumidifiers; Growing plants which absorb humidity, colder and shorter showers, as well as a home dehumidifier, are all ways to effectively reduce humidity.

By staying in the range of 40-70%, you will not only be able to decrease the risk of coming down with an infection but also reduce the chances of spreading and affecting others around you!
(Arundel, Sterling, Biggin & Sterling, 1986)

Temperature

Nasal patency is a measurement of how large the opening of your nostrils is, Temperature can play a large part in affecting it. An increase and decrease in environmental temperature will affect your nasal patency. A study from the Department of Oto-Rhino-Laryngology University Hospital shows that when exposed to a warmed/heated environment (40 °C), your nasal patency increases. Increasing air airflow into the nose, soothing the otherwise blocked nose. This is contrasted when exposed to a cold environment (6 °C), where your nasal patency will decrease. Thus, making your nose more congested. It is noted that an increase or decrease in environmental temperature will affect the blood flow in the nose (a reference to chapter 2 on 'swelling of blood vessels'). However, the swelling if any experienced will not be the cause in the decrease in nasal patency. Instead, the change in temperature is the main cause that directly links temperature to nasal patency. (Olsson, 2005)

Temperature also creates an effect on increasing nasal mucus velocity through methods of nasal inhalation of hot water vapour. (Saketkhoo, Januszkiewicz & Sackner, 1978)

Air Pollution/Particulate Matter

Air pollution constitutes the presence of chemicals in the air, including sulphur and nitrogen dioxide, particulate matter as well as industrial chemicals. Nerve receptors are found very close to the surface of the nose and mouth, only separated by a thin water membrane. As such, volatile irritants from the environment can penetrate and react with the nerve receptors in the membrane of the nose and mouth. These nerve receptors react to these airborne chemicals, causing inflammation through the release of neuropeptide chemicals. The neuropeptides result in a cascade of events, causing further inflammation, swelling and sometimes pain. Inflammation in your nostrils will contribute to/amplify the formation and severity of a blocked nose. (Sullivan, Van Ert, Krieger & Peterson, 2013)

Haze

For every increment of 10PSI from the daily mean, there averages a 0.7% increase in upper respiratory symptoms. (Dockery & Pope, 1994) This tells us that in 2015 when Singapore was reported to have a PSI of 152 (Khew, 2015), about 115 points higher than the mean of 37. This caused a minimum 8.05% increase in the presence of symptoms such as discomfort in nasal passages, blocked or runny nose and excess mucus.

Olfactory inhibition

The presence of air pollutants also affects one's ability to smell as effectively. (Guarneros, Hummel, Martínez-Gómez & Hudson, 2009) When comparing two cities, Mexico City, an area with high levels of pollution, versus Tlaxcala separated from Mexico City by a mountain range, an area with low levels of pollution. 6 out of 30 test subjects from Mexico City exhibited a smelling score which were at Hyposmic levels, a range of impaired olfactory ability. Whereas none of Tlaxcala's test subjects had a score in this category, all being able to smell scents at a normal level.

This gives rise to the conclusion that high urban air pollution levels do affect otherwise healthy individuals, causing substantial damage to olfactory senses. Public health concerns include what age symptoms show and until what age is air-pollution-hyposmia is reversible, to combat loss of smell. (Sullivan, Van Ert, Krieger, Peterson, 2013)

CHAPTER 5: HOME-MADE SOLUTIONS

This chapter is about the testing of home-made remedies. The solutions have been derived from a range of journal articles, blog posts and rumours. We have referenced them with the speculated theory behind each remedy and personally tested their effectiveness! We have given each solution a rating of out '10', '0' denoting total ineffectiveness and '10' representing a very effective solution. Following that, we have given our recommendations for you to integrate this solution into your life!

Using a Neti Pot

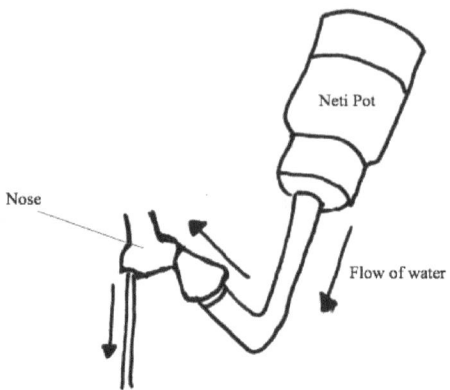

The theory behind using the neti pot is to directly administer saline (saltwater) into the nostrils, as a form of nasal irrigation. The solution will be poured into one nostril and will flow out the other as can be seen by the illustration. This dilutes and softens mucus, which aids in the removal of congested mucus and creates a healthy humid environment inside the nose. This will remove and prevent bacterial build up in the nose as well as clear the congestion.

Test: 10.30am Friday morning, while working on the book. Neti Pot and 0.9% m/v Sodium Chloride Solution by Baxter was used in this experiment.

Following the instructions, I added 300ml of the Saline solution into the Neti Pot and attached the 'Adult' Nozzle Head. With my head slanted to the side, I stuck the nozzle into the nostril at a higher elevation and administered the solution. It felt slightly uncomfortable at first, but the water flowed up my nostril and out of the other. After the flow became consistent, I changed sides, tilting my head the other way and putting the nozzle into the other nostril. The mucus in my nose softened up from being diluted and was very easily blown out. Following that, my nose was completely clean and comfortable. The whole process felt somewhat therapeutic. The full extent of the effects lasted for more than an hour and my nose remained very comfortable for the rest of the day.

Rating: 9/10

Some drawbacks include that you have a sink to perform the process but, more importantly, you have to purchase a neti pot and the saline solution/mix the solution yourself. However, after you obtain both, the process is simple and effective. It was fast, fuss-free and can be done anywhere, as long as there's a sink.

This solution can be used regularly and has minimal major drawbacks, it is highly recommended.

Drinking hot beverages

The theory behind consuming hot beverages was to induce steam/water vapour inhalation and to bring up the temperature of the nose. The increased humidity will reduce the viscosity of the mucus, allowing it to flow and increased temperature will increase nasal patency. As this exposure is more indirect, the period of lasting positive effects is not expected to last as long as other more direct solutions, but it certainly worth a shot. (Sanu & Eccles, 2009)

Test: 10.30am Monday Morning, while working on the book. Using hot boiled water cooled to 60 °C

After 6 minutes of sipping the hot water and inhaling the water vapour, my mucus got much runnier and soft. This allowed me to clear it effectively. The effects lasted as I was drinking the water and remained for 12 minutes. After that, the mucus started returning into a thickened state, but it was still clearer than before.

Rating: 6.5/10

The duration of hot water's effects was short lasted and as a solution, it is only mediocre. That being said, it is inexpensive, extremely convenient and doubles as a good health practice to staying hydrated.

It is our recommendation to consistently sip or drink hot water periodically while suffering from nasal congestion. This will produce more consistent positive effects and some much-needed relief throughout your workday!

Eating instant noodles

It may seem far fetched, but there was some thought put into it at first. First is using humidity; while the noodles are cooking and as you eat it, water vapour accumulates. As the noodles are within proximity to the nose, humidity is indirectly administered into the nostrils. Second is high temperature; As the noodles are cooked with boiling water, upon consumption, the noodles and steam are hot. The high temperature from the steam increases the temperature of your nose. Theoretically, this should increase nasal patency, dilute congested mucus and increase overall comfort within the nose. It is even hypothesized that scent or taste-related receptor activation may also produce a positive effect in increasing mucus velocity. (Saketkhoo, Januszkiewicz & Sackner, 1978)

Test: 9.30pm Sunday Evening, while working on the book. NISSIN "Chicken" Flavour Instant Noodles were used in this experiment.

Within 3 minutes, the viscosity of my mucus greatly reduced. This allowed the mucus to be easily blown out and cleared. For the following 6 minutes, as I continued consuming the noodles and soup, my nose had remained clear. The heat of the noodles also increased nasal patency in both nostrils, allowing me to breathe more comfortably. The full extent of the positive effects of the noodles continued for >45 minutes after finishing it and lasted throughout the rest of my evening.

Rating: 7.6/10

The reason behind this rating is that it may not always be the best time to be eating instant noodles, depending on location (e.g. office housekeeping rules). Another drawback is instant noodles are not very healthy, especially not for regular consumption. However, the noodles were very effective. The congestion cleared very fast and the effects were long-lasting. Plus, the noodles double as a supper or a snack to refuel and enjoy. It was also noted that the waiting time for the noodles to cook was torturous.

We recommend that it is a great solution for nasal congestion during meal times, but not for every occasion.

Taking a hot shower

The idea of taking a hot shower based on the intention of increasing the temperature of the surroundings, the temperature of your nose and increasing the humidity temporarily. This should increase nasal patency and help prevent sinus infection in the long run.

Test: 12.20pm Thursday Afternoon, while working on the book. The temperature of water used in a hot shower is < 41 °C

After being in an air-conditioned room for a while, where the humidity is lower than the recommended range and where the temperature is cold, my nose started running and was congested. The hot shower lasted 6 minutes and 50 seconds, the nose was starting to clear, but my nose remained uncomfortable. 3 minutes after the shower, my nose was fully clear. The congested mucus was softened and removed by the moisture from the shower and the increase in nasal patency also made breathing easier. The full extent of the effects lasted for more than 40 minutes after the hot shower and my nose stayed relatively comfortable for the rest of the afternoon.

Rating: 7.9/10

The reason behind this rating is that it is not always convenient to take a hot shower, depending on time and location. That being said, the shower was simple and effective, as well and long-lasting. It improved overall comfort and felt good.

We recommend that this is a perfect solution for nasal congestion in the morning or at night, where you can take a hot shower before leaving home or before going to bed. This way your nose will be decongested for a congestion-free day and/or for a more comfortable rest at night. This can be a regular strategy to keep your nose free from congestion and in good condition.

Eating spicy food (Capsaicin)

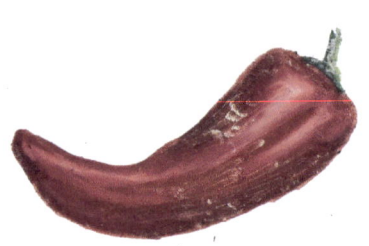

The theory behind eating spicy food is that the peppers used to provide the spice have capsaicin. Capsaicin is the active ingredient in chilli peppers, which usually acts as an irritant for mammals. It produces a burning and sometimes numbing sensation with any tissue it comes into contact with.

Capsaicin very readily activates the nasal receptors, exciting the neurons and causing the release of neuropeptides. This results in two benefits to your nose; Chronologically first, these neuropeptides act as signals which trigger an inflammatory response. This causes temporary excess mucus production, combined with condensation from an external source, it can cause your nose to "overflow". The stoppage of the overflow after a while will coincide with the clearance of your nose.

After being exposed to capsaicin for a long time, nasal receptors may also be overstimulated. These neurons get "tired" and are no longer able to respond to a wide range of irritants and stimuli. This causes your nose to be less sensitive for a long time and reduce vasomotor rhinitis symptoms to persist. (Fokkens, Hellings & Segboer, 2016)

Test 3.10pm Friday Afternoon, Lunch break. The spicy food we used for this experiment was a Mala Xiang Guo (麻辣香锅) at a moderately spicy level (中辣).

The bowl of Mala we ate for the experiment!

Within 3 minutes of eating, the viscosity of the mucus in my nose had dropped tremendously, almost to even a watery state. It was easy to blow out, with little force, and fully clear my nose. This was a result of the water vapour from the food, softening the excess mucus produced from capsaicin stimulation. The full extent of the positive effects lasted for more than 45 minutes and my nose remained clear for the rest of the day, this is indicative of my nasal receptors becoming desensitized and not reacting to other stimuli. The heat from consuming the food also increased nasal patency, which aided in making 'breathing' through my nose easier.

Rating: 7.4/10

Notable drawbacks would include that not everyone enjoys the heat of spicy food, for non-spice lovers this solution may be torturous. On a hot day, eating spicy food may not be a good idea and as this is an edible solution, it may also not be convenient to eat whenever and wherever. However, capsaicin works and is very effective. Even on a bad nose day, the spicy food produced a totally clear nose and instant relief. Doubling as a meal, it can be pretty convenient too.

We recommend that this solution is used on days where you are experiencing a particularly badly congested nose, the capsaicin will act as an extra boost to clear out the nose!

CITATIONS

Mygind, N. & Änggård, A. (1984), Anatomy and physiology of the nose—pathophysiologic alterations in allergic rhinitis. *Clin Rev Allergy*, 2(3), 173-188. https://doi.org/10.1007/BF02991098

Kahana-Zweig, R., Geva-Sagiv, M., Weissbrod, A., Secundo, L., Soroker, N., & Sobel, N. (2016). Measuring and Characterizing the Human Nasal Cycle. PloS one, 11(10), e0162918. doi:10.1371/journal.pone.0162918

Corey, J. P., Houser S. M. & Ng, B. A. (2000) Nasal Congestion: A Review of its Etiology, Evaluation, and Treatment. Ear, Nose & throat journal, 690-702. https://www.researchgate.net/publication/12311196_Nasal_Congestion_A_Review_of_its_Etiology_Evaluation_and_Treatment

Stewart G. A., Peden D. B. ,Thompson P. J. & Ludwig M. (2012) 5 - Allergens and air pollutants. Allergy (Fourth Edition), 73-128. https://doi.org/10.1016/B978-0-7234-3658-4.00009-3.

Ferrándiz, M., Martín-Galiano, A. J., Schvartzman, J. B. & de la Campa, A. G. (2010), The genome of *Streptococcus pneumoniae* is organized in topology-reacting gene clusters. *Nucleic Acids Research*, 38(11), 3570–3581, https://doi.org/10.1093/nar/gkq106

Bourouiba, L., Dehandschoewercker, E., & Bush, J. (2014). Violent expiratory events: On coughing and sneezing.

Journal of Fluid Mechanics, 745, 537-563. doi:10.1017/jfm.2014.88

Padrid, P., & Church, D. B. (2008) Drugs used in the management of respiratory diseases. *Small Animal Clinical Pharmacology (Second Edition)*, 18, 458-468. https://doi.org/10.1016/B978-070202858-8.50020-8.

Quraishi, M. S., Jones, N. S. and Mason, J. (1998), The rheology of nasal mucus: a review. *Clinical Otolaryngology & Allied Sciences*, 23, 403-413. doi:10.1046/j.1365-2273.1998.00172.x

Eylers, J.P. (2001), Mucus and Slime: Structure and Rheology of Natural Polysaccharides. *Encyclopedia of Materials: Science and Technology*, 5848-5850. https://doi.org/10.1016/B0-08-043152-6/01018-4.

Carrington, S. D., Clyne, M., Reid, C. J., FitzPatrick, E., Corfield, A. P. (2010), Microbial interaction with mucus and mucins. *Microbial Glycobiology*, 33, 655-671 https://doi.org/10.1016/B978-0-12-374546-0.00033-X.

What the colour of snot really means. (2017, June 28) Retrieved from: https://health.clevelandclinic.org/what-the-color-of-your-snot-really-means/

Pappas, D. E. (2018). 26 - The Common Cold. Principles and Practice of Pediatric Infectious Diseases (Fifth Edition), Pages 199-202.e1, https://doi.org/10.1016/B978-0-323-40181-4.00026-8.

Arundel, A. V., Sterling, E. M., Biggin, J. H., & Sterling, T. D. (1986). Indirect health effects of relative humidity in indoor environments. Environmental Health Perspectives, Vol. 65, pp. 351-361, https://ehp.niehs.nih.gov/doi/10.1289/ehp.8665351

Olsson, P. (2005). Studies of Blood Flow in Human Nasal Mucosa with 133 Xe Washout Technique and Laser Doppler

Fiowmetry. Pages 6-30. Retrieved from: https://www.semanticscholar.org/paper/Studies-of-Blood-Flow-in-Human-Nasal-Mucosa-with-Xe-Olsson/71ec75e958e8e8b536b5c8c01806af285f9b2329

Sullivan, J. B., Van Ert, M. D., Krieger, G. R. & Peterson, M. E. (2013). Chapter 14 - Indoor Environmental Quality and Health. Small Animal Toxicology (Third Edition), Pages 139-158, https://doi.org/10.1016/B978-1-4557-0717-1.00014-4

Khew, C. (2015, October 25). Haze: One-hour PM2.5 reading soars to 442. Retrieved from: https://www.straitstimes.com/singapore/haze-one-hour-pm25-reading-soars-to-442

Dockery, D. W., Pope, C. A. (1994). Acute Respiratory Effects of Particulate Air Pollution. *Annual Review of Public Health*, Vol. 15, 107-132. https://doi.org/10.1146/annurev.pu.15.050194.000543

Guarneros, M., Hummel, T., Martínez-Gómez, M., Hudson, R. (2009), Mexico City Air Pollution Adversely Affects Olfactory Function and Intranasal Trigeminal Sensitivity, *Chemical Senses*, Vol.34(9), 819–826, https://doi.org/10.1093/chemse/bjp071

Sullivan, J. B., Van Ert, M. D., Krieger, G. R., Peterson, M. E. (2013), Indoor Environmental Quality and Health. *Small Animal Toxicology (Third Edition)*, 139-158. https://doi.org/10.1016/B978-1-4557-0717-1.00014-4.

Sanu, A. & Eccles, R. (2009). The effects of a hot drink on nasal airflow and symptoms of common cold and flu. Rhinology. 46. 271-5. https://www.researchgate.net/publication/23790050_The_effects_of_a_hot_drink_on__nasal_airflow_and_symptoms_of_common_cold_and_flu

Saketkhoo, K., Januszkiewicz, A., Sackner, M. A. (1978).

Effects of Drinking Hot Water, Cold Water, and Chicken Soup on Nasal Mucus Velocity and Nasal Airflow Resistance, Chest, Volume 74, Issue 4, Pages 408-410. https://doi.org/10.1016/S0012-3692(15)37387-6.

Fokkens, W., Hellings, P., & Segboer, C. (2016). Capsaicin for Rhinitis. *Current allergy and asthma reports*, *16*(8), 60. doi:10.1007/s11882-016-0638-1

Just being nosy is done by:

Authors: Goh Zhi Han, Ambrose & Chew Ming En, Lucas
Cover Page Artist: Tan Reagan
Illustrator: Chew May-Anne
Facilitator: Zhang Peng Chi

ABOUT THE WRITERS

Lucas (Left) and Ambrose (Right)

We are 2 second-year students studying in Temasek Polytechnic's Biotechnology course. The inspiration for writing this book comes with a shared history or experiencing nose problems and daily nasal congestion. The writing of this book has certainly opened our eyes to how much time, effort and commitment it takes to fully construct a book and publish. We are grateful for the opportunity to work on this project as a form of 'Guided Learning'.

We would also like to express special thanks to Reagan Tan, the illustrator of the book cover, May-Anne Chew, for helping to illustrate the drawings in the book and Dr Zhang Peng Chi, our Guided Learning Facilitator who sparked the idea of book writing and has guided us through the entire process with his past experiences and invaluable suggestions!